When The Grid Goes Down

Disaster Preparations and Survival Gear
for Making Your Home Self-Reliant

Tony Nester

When the Grid Goes Down
Disaster Preparations and Survival Gear For Making Your Home Self-Reliant
By Tony Nester

Copyright December 2012 by Tony Nester

Published by Diamond Creek Press
PO Box 2068
Flagstaff, AZ 86003
1-928-526-2552
www.diamondcreekpress.com
Email: info@urbanskills.net

Cover photo by Michael Bielecki, MissingFramePhotography.com

ISBN 978-0-9713811-4-8

Also by Tony Nester:
Practical Survival Tips, Tricks, & Skills
Desert Survival Tips, Tricks, & Skills
The Modern Hunter-Gatherer: A Practical Guide to Living Off the Land
Desert Survival DVD
Tony Nester's Practical Urban Survival DVD Series, Volumes 1 & 2

Disclaimer
The information provided in this book is for academic purposes only. The author
made every effort to ensure the information was correct at the time of publication
and the author, Diamond Creek Press and Ancient Pathways, LLC assume no liabil-
ity for any laws broken by the reader nor any personal injury or injury to anyone
else for the use or misuse of the information contained in this book.

Visit us at www.apathways.com for information on our other books or survival
training courses.

Table of Contents

Acknowledgements

Once again, thanks to friend Randy Miller who has lived and breathed self-reliance before the concept was mainstream and who has always graciously shared his wealth of insights and experience with the lifestyle. To my good friend and fellow survival instructor Mike Masek who has been there through thick and thin and always lends invaluable lessons from his globetrotting days in remote regions. I also want to acknowledge Jerry Ward of Ozark Mountain Preparedness with whom I've enjoyed many fine conversations in both the urban and wilderness worlds. Thanks to WEMT instructor and guide extraordinaire, Bill Downey who has lent his considerable remote medicine experiences to my knowledge of disaster first-aid and backcountry medical skills. Thanks to Gabe Suarez and Tom Sotis, two legendary pioneers in the field of combatives who have been instrumental in shaping my thinking on personal defense. Lastly, to the many first-responders and real-world survivors who have shared their stories and insights into what it takes to prevail and how community effort reigns supreme when the chips are down.

Preface

I make my living as a full time survival instructor and have been teaching courses in bushcraft and survival since 1989. Over the years, students of mine have often asked "What's the difference between urban and wilderness survival?" Not much is my response. Your body doesn't care when you have hypothermia, broken ribs, or dehydration if it's in the wilds or downtown Seattle. We humans have certain physical priorities that must be taken care of or we will die.

The methods for meeting your priorities are different but the priorities themselves remain the same regardless of environment. The methods for surviving in the backcountry will entail treating medical injuries, building a shelter for warmth, staying hydrated by locating a water source, and communicating with searchers. In an urban disaster, you will also be tending to medical issues, staying warm and dry (shelter), tapping your water reserves and trying to communicate with family. There's quite a bit of overlap.

One difference between the city and the wilderness is in the area of self-protection. I don't encounter many roving bands of hungry bandits on my day hikes in the mountains. I am very likely to in a depressed urban catastrophe where stores lay bare for weeks. Desperation can happen after day three in a large city when the grocery shelves empty and the water tap is dry. By day five, anarchy and roaming groups of hungry souls will be looking for those who can provide, peacefully or otherwise. This is another reason to keep your preps low-profile.

In addition to my work in the survival field and the lifestyle of self-reliance my family and I have worked towards, I've also interviewed survivors in natural disasters like the Indonesian tsunami, first responders in Katrina and soldiers deployed to war-torn regions where catastrophes seem to be a daily way of life. All have mentioned the same thing: food, water, and personal safety are what life boils down to in an urban disaster. Little else matters when life is driven into a corner.

When not out in the field or on the road teaching, I live in a passive-solar, strawbale house in the mountains with my family. We garden, forage, hunt and have incorporated many of the strategies in this book into our daily lives but it didn't happen overnight. Self-reliance is something that you ease into working at each piece of the puzzle until, one day, the pieces fit together a little smoother and you become less dependent on the tether back to the city. It's something we still build upon each year. Throughout these pages, I will be drawing upon the methods we have used to build a more personally sustainable lifestyle.

For years, my company also ran a 9-week program called the Southwest Semester Program in traditional skills and bushcraft in northern Arizona. My students and I came to live primitively, in the backcountry for over two months and learned to wrest a living from the wilds using the old skills. While part of this time was spent on numerous walkabouts in the desert and field trips to tribal lands, we all lived off-grid at our basecamp learning the age-old skills of self-reliance. Our basecamp is spread over 40 acres of high-desert that backs up to miles of rugged wilderness. In this remote setting, we lived, worked, ate, slept, and took care of life's most basic necessities while learning to make do with less dependence on the modern system.

Taking care of a small tribe of individuals for over two months provided great insight into needs vs. wants in a setting that was light on modern conveniences. How much food, water requirements and storage, firewood quantities, shelter material, hand-tool types and maintenance, cooking gear, laundry materials, repair items, clothing and laundry issues, pest-control (most were eaten!), sanitation and medical gear to name a few items, were all worked out via trial and error and further refined with each program. The weather, from torrential rain and snow to triple-digit heat was also an excellent teacher that operated on a pass/fail system and provided instant feedback.

There is certainly a long list of skills and topics that you can tackle when delving into a self-reliant home. A survival gear industry has sprung up in response to this and unfortunately fosters a doom & gloom mentality to peddle their goods. You will see that time and again, that low-tech methods and gear are often the best ones to have in place. To the person just getting started the gear lists and survival methods can be overwhelming and I have kept to just the six core areas for preparing. Obviously you can expand on this by getting into solar energy, wind turbines, livestock and poultry, gardening and so on but, as with my other books, this one sticks to the basics and is designed to provide the foundation skills and tips.

Think of a long-term grid-down situation as basically camping out in your home because that is what you will be doing. Sleeping under a pile of blankets, dressing in layers- heating the body not the house, cooking on propane or a campfire, gathering intel from the radio (and possibly internet?), reading by candlelight or headlamp, skipping the daily shower in favor of a bi-weekly sponge bath, collecting rainwater and being your own first-responder- this will be the new norm if a disaster cripples your region.

The more planning you have in place for the above areas, the less vulnerable you will be when the grid gets disrupted. Severing dependency upon the system need not be abrupt or even done 100%. Begin by slowly building a more self-reliant lifestyle using the methods outlined in this book, each month focusing on a specific area.

Eventually, you will find that your growing confidence and know-how will enable you to prevail regardless of what crisis hits. That being prepared is a lifestyle choice and not simply a reaction to an economic nosedive or another seasonal hurricane or ice-storm. The essence of this book is all about getting back to the skills and independent mind-set we had in this country a few generations ago.

Best wishes on your journey towards greater freedom and security.

Tony Nester
Flagstaff, AZ
2012

HOW TO CREATE A SELF-RELIANT HOME

The Six Key Survival Priorities
1. Water
2. Food
3. First-Aid
4. Home Security and Personal Defense
5. Heating, Cooling, & Energy Needs
6. Hygiene & Sanitation

Water

Despite what they show in the heavily-scripted reality shows on survival, you can't live long without water and also can't magically extract it from the ground using a solar still. Water is at the top of this list for the simple fact that we humans must stay hydrated in order to survive. Whether you are a triathlete, Green Beret, or live in a desert cave, you can't condition your body to go without this precious substance.

Having spoken with many survivors who were struck by disasters in Japan, Thailand, and India, as well as the debacle of Katrina, all of them recounted

stories about the lack of fresh drinking water and mile-long resupply lines at aid stations. Most spoke about how they could handle the other priorities such as improvising shelter items from scavenged materials (clothes, blankets, furniture) and banded together to provide medical assistance and share food. However, fresh water was scarce and dehydration endlessly tugged at their bodies and minds. This is where having, at the minimum, two gallons per person, per day of water in your house will pay off.

Food

Urban Survivors under real-world conditions have gone for days without food and wilderness survivors have gone 42+ days. Certainly our bodies are hard wired for a feast or famine existence from our genetic past as hunter-gatherers but the prepared survivor has food reserves in place at home and emergency rations in their bail-out kit.

Life will be physically stressful enough when a disaster strikes and in the recovery phase that follows so having a stocked pantry is an essential area of preparation. By stocked pantry I don't mean a three-month supply of MREs but rather items that are not too far removed from what you already eat on a regular basis.

I also wouldn't plan on thinking you will head to the hills and live off the land. If you are versed in such hunting & gathering skills then you will certainly

have an edge during the lean times but wild game procurement should be used for augmenting your existing stores and not as a replacement for being prepared.

First-Aid

When the infrastructure is damaged during a large-scale disaster or severe weather hampers 911 response times, you may become your family's own first-responder. This is where a quality first-aid kit comes into play. One that is designed with remote medical issues in mind. I would also highly recommend taking a 2-day Wilderness First-Aid or 9-day Wilderness First Responder class. These are the types of hands-on, remote medical skills that you will find most useful when faced with an emergency where help is days away. If you are a parent, you will be grateful to have such training under your belt regardless of whether an urban disaster ever befalls you.

Home Security and Personal Defense

Personal security measures for you and your family along with your home are going to vary depending on budget, geographic location and gun/weapon laws, and the size of your house. Home defense and security become paramount in a long-term grid-down situation regardless of whether you live in the city or the country. A few days without food, water, and warmth will peel back the veneer of civilization and reveal a side to people that you wouldn't believe. Smaller communities where the social fabric and relations are tighter usually fare better in my experience.

Though I am an advocate of firearms and combatives training, home security entails more than having a 9mm in your bedroom or being versed in the martial arts. Safety measures begin with the outside of your house such as having proper lighting, deadbolts, solid-core doors, perhaps a wireless alarm system, and, of course, plenty of situational awareness. The interior entails having a family escape plan, firearms or defensive tools, possibly a dog, a safe room, and so on. As with the other priorities, home security should involve a layered approach.

Heating, Cooling, & Energy Needs

With both short and long-term disasters the grid may be down for an extended period. Maybe it won't be a problem going without power for a day or two depending on where you live. However, what if you have an elderly family member whose medical devices require electricity or you simply need to get some news or communicate with others? Backup and low-tech energy systems have come a long way in the last ten years due to the outdoor and RV industry. Portable chargers, generators and solar panel units are now available that will enable you to charge your laptop, preserve and cook food, stay warm or cool, obtain weather reports and read at night. Having a few low-tech devices will also enable you to weather out the heat and cold.

Hygiene & Sanitation

Where will the human waste and trash go once services stop? How will you bathe and clean dishes? Our large metropolitan areas revolve around a tremendous infrastructure of workers and services to maintain simple things like trash management and waste water. When the power is out for more than a few days,

you will need to adopt the approaches used by the backpacking and outdoor industry as I will discuss in the pages ahead.

This is a key area to tackle in your home self-reliance plans as a breakdown in hygiene within your family can spell disaster in another fashion through a visit from the stomach bug. We have seen this on long wilderness trips, where it only takes one person in the group failing to maintain handwashing, and then the entire group is wiped out by diarrhea or stomach ailments a few days later. Under the third world conditions that will develop after an urban disaster, hygiene will rise to prominence. Daily rituals involving constant hand-washing or using hand-sanitizer along with bleaching down counters and food prep areas will be essential to physical health in your home.

When to Stay and When to Bail Out?

Often I am asked this question in my urban survival classes. What determines when to stay put in your fortress of solitude and when you should split with your Bail-Out Bag and essentials? Each situation is different and you

will want to use the Six Priorities as your gauge, especially with regards to a long-term disaster.

Obviously if a CAT 4 Hurricane or major wildfire is headed your way, then evacuation is in order. For long-term grid-down scenarios though, there are some other things to ponder. If your water is nearly out and there is no chance of resupply then that is a pressing issue as you won't last long. Are home and neighborhood defensive barriers beyond your control as gangs move in with plenty of firepower? Is your child's asthma medication running low and the local pharmacies shuttered? Are you looking at your home becoming an icebox as winter approaches and your furnace no longer operational due to lack of power?

The Priorities are labeled that for a reason. If you can no longer take care of them in a long-term situation, then it may be time to consider bailing out to a relative's or friend's location or your retreat.

There are really four plans to develop:

1. Home self-reliance and the Six Priorities outlined herein
2. Evacuation strategies and your Bail-Out Bag
3. Strategies for life on the road (bailing out) as you get from your home to the new location
4. Re-establishing the Six Priorities at your new location

At some point, you will want to sit down and discuss this in detail with your family or like-minded friends. I have detailed number two above in my previous book, *Surviving a Disaster*. Lastly, once you have gathered all the information from the media and assessed the situation, remember to listen to your gut feelings. Each situation and every city's evacuation issues are going to be different and I can't give you a cookie-cutter answer on the exact moment to bail out.

As I see it, if you head out of the city prior to what looks like an impending disaster and find out the next morning on the news that your departure was uncalled for, then at most, you are looking at an overnight stay at a friend's house or in a hotel, a practice run of your evacuation strategies, and then heading back home. But if you wait and you don't flee as the dark clouds are rolling in…

How Prepared is *Being Prepared*?

Reading some of the advice on various urban survival forums and looking at the survival industry, you could draw the conclusion that you must have a remote mountain bunker or private island that can house and feed your family in isolation for five years while the world unravels. So where does the average person start and what should they prepare for exactly?

Being "prepared" means different things to each person and will reflect your geographic location, family size, budget, house size, regional weather, and whether you live in the city or the country. Let's look at the overlap and differences between short and long-term disasters as each involves a different level of planning and preparedness though the key priorities remain the same.

Short-Term Scenarios

A blizzard, wildfire, ice-storm, flood, or chemical spill on the railroad tracks would fall into this category. Normalcy and a return to municipal services will occur in a few days to a week. At the very least, you should have enough supplies for a 7-day grid-down situation in your home and city with enough food, water, and medical supplies to take of yourself beyond the much-touted 72-hour federal agency recommendations. Hunkering down in place or evacuating to a friend's house across town is all that is in order as life will return to its usual pace shortly.

Long-Term Scenario

Here, we are looking at a city or large-scale disaster caused by a hurricane, terrorist attack, a global pandemic, or a nuclear reactor catastrophe that could devastate a region for several months (or a year or more as with an avian pandemic). These scenarios are a world apart in their impact from being holed up at home for two days during a freak, winter snowstorm.

This is where advanced planning along with preparation and an investment in food, water storage, and camping gear will allow you and your family to establish a sense of relative normalcy. Here we aren't talking about surviving so much as living and maintaining continuity in your daily life. There may be tough times for sure, but your network of like-minded family and friends, food and water replenishment systems (gardening, hunting, fishing, etc…), local natural resources, bartering, and black market capabilities will enable you to ride this out into whatever the region or world transitions next.

The key here is that you have a head start compared with those who are unprepared and have to begin with the core priorities from scratch while coping with environmental and economic duress.

The difference between short term and long-term isn't so much in the priorities as with both you will still need to eat, stay hydrated, handle medical issues, and so on. The difference is that with the short-term you can resupply after a few days and see a return to normal services. With the long-term, there may be sporadic resupply or none at all. You may be solely responsible for replenishing food and medical supplies as grocery shelves empty and supplies dwindle due to disruptions in the transportation network and city resources. In this scenario, even a year's supply of food and resources may not be enough.

This is where knowledge and experience with gardening, hunting, fishing, foraging, herbal medicine, and wilderness first-responder skills come into play. I highly doubt we will all be walking around in buckskin with primitive bows. Knowing how to feed your family, stay warm and dry, quench your thirst, and tend to medical issues are life priorities that haven't changed from the time of our ancient ancestors right up through the modern world.

My advice is to start with the Six Priorities to handle, at the very least, a short-term situation and then build from there upon your lifestyle while acquiring skills and information from others versed in long-term skills.

Solo vs. Group Survival

Solo survival is always grueling and the hardships, both physical and psychological, become amplified when you are alone compared with surviving in a group. Humans have survived this long because of culture! We are social animals and a group has a far better chance of surviving than the lone individual. For one thing, the more brainpower and brawn, the better. The drawback is that you need more food, water and resources.

I've done both solo survival outings and group survival trips where we went out as a small hunter-gatherer "band." These outings ranged up to a month and with very little modern gear. Psychologically, being in a group was far more pleasant and easier, with our daily foraging trips and camaraderie around the evening campfire. Not to mention how much more wild food we were able to procure than if I had been alone. Simply from a caloric-expenditure perspective, a small group (6-8 people) of trained individuals, in a good environment and at the right time of year, will fare better than the solo survivor.

Yes, many, many people have survived solo. Stories abound but the loneliness and despair was worse than if they had been able to commiserate with and receive support from others. Recall the Andes plane-crash survivors in the 1970s. They were a group of young athletes who were already a team when they crashed. Their courage and will to live was fortified by their bonds and mutual support.

The Layering System

The Navy SEALs have a saying, that "two is one and one is none." This refers to redundancy in critical gear. Throughout this book, you find repeated references to this layering approach as one of anything critical is a poor system. For instance, if you have a bail-out kit at home and you are stuck in a grid-down situation at work, all your preps aren't going to help. The layering approach has you equipped with a bail-out kit in your home, your vehicle, and your office so all the bases are covered.

The same applies with water storage. You have two dozen bottles of water in the laundry room, followed by several 7-gallon water jugs in the garage, and then a 55-gallon rain barrel connected to the back gutter. You also know how to purify water from the nearby stream and have all of the local springs mapped out for your region. Let the layering approach guide you in each of the six key areas of home self-reliance.

Water Storage and Purification Methods

You can't live long without this precious substance. Humans have gone weeks without food under survival conditions but without water, your shelf-life is limited to days and possibly even hours if the heat is extreme.

Consider for a moment what would happen to residents in Phoenix or Las Vegas if there was a blackout lasting several weeks due to an incident that cripples the grid. During the summer, it can spike past 115 degrees F in the afternoon. Working and living in a post-disaster setting like this would cause a person to consume two or more gallons of water per day. Life without water in these unforgiving city-ovens can be limited to hours if the proper precautions and water planning aren't in place.

The general rule I recommend, and that we have found useful at our house, is to have two gallons of water per person, per day on hand. This is just for consumption and cooking not for dish washing, laundry, livestock, or the garden.

For a family of four, this will come out to eight gallons per day and 240 gallons a month. That's a lot of water bottles to buy up in pallets, eh? As noted, I am a real fan of layering your critical survival items so that your water is broken down into several storage systems from large to small.

Water Storage Options

Here's what we've done. We have a 210-gallon water tanker that collects rainwater off our roof. This is the main vein for general use like dish washing, animals, and the garden along with just being an emergency back-up system. The next step down are two, 55-gallon poly-barrels (see photo) with accompanying hand-siphons. These blue barrels can be found at feed stores and some big-box pet stores for around $25. Some of these come from the commercial restaurant industry, so be sure to get the ones other than those used for storing garlic or olive oil—something I discovered the hard way.

Next, we have six of the blue 7-gallon cubes behind our shed (north-facing). These can be found in camping stores or at WalMart. I use these on my field-courses and they last for about eight months of punishment before the corners crack from the constant exposure to UV rays (at 7000 feet where I live that's a lot) and daily handling. If you are storing these in your garage or basement out of the sunlight, then you should be fine. At the bare minimum, get two of these

blue cubes and you will have 14 gallons of water which will last one person seven days or a family of four for three days. Consider this a place to start and then you can build your water stores up from there.

Lastly, we have 1-gallon containers in the form of iced-tea bottles. We have a few of these on hand around the property and one in each vehicle as this is a convenient size to tote around.

One of my students in New York City has 30, of these 1-gallon jugs of water stowed in his small apartment (he is a hardcore apple-cider addict). At two gallons per person, per day, he has enough for nearly two weeks.

He cycles through these each month which is what I recommend with all

stored water. We have used all of the previously mentioned water storage systems long-term and have never had any bacteria build-up as long as we are constantly rotating the water. Keeping containers out of the sun is helpful. Every three months after draining a water cube or the big barrels, I will completely bleach out the innards, rinse and refill. Bleach is a must-have item for maintaining your water barrels long-term and for water purification, as I'll discuss next.

> The three enemies of long-term water storage are: UV, heat, and exposure to the air (dust, microscopic particles getting in). Reducing these three will reduce bacteria buildup. This is where bleach comes in along with having your water cache out of the direct sunlight and intense heat. If you can't do this, then you need to rotate your water frequently.

Water Purification Methods

In a crisis, waterborne diseases are going to be a major concern so avoid the reality-TV theatrics of gritting-your-teeth-to-strain-out-the-big-stuff and equip your home with some of the following low-tech items.

These four methods of purifying water can be used for treating contaminated sources in both urban and wilderness settings.

Boiling

Boiling kills both viruses and bacteria but does not remove chemicals such as fuel, lead, or other toxins that may have leached into the city's water lines when infrastructure damage occurs. According to a CDC researcher I spoke with, you need to boil water for only 1 minute to kill the micro-organisms that are present. Actually, it's 160 degrees F to be precise but since we humans can't tell when water is at that temperature on the stove or campfire, the CDC recommends the visual of the rolling boil to know you have exceeded the necessary temperature.

UV Treatment

A SteriPen zaps the water with UV rays and kills both bacteria and viruses. Insert the tip of the pen into a liter of clear water and turn on the switch for 60 seconds. The drawback is that you need clear water to begin with. In murky or silty water, the UV rays won't penetrate to the depth of the bottle and treatment will be incomplete. SteriPens are great for an urban survival home kit but it does require four AA batteries, won't function in extreme cold and is delicate. I keep mine in a bubble-wrap sheath when not in use. The company also makes a version that has a solar charger. I use a SteriPen when I travel internationally for treating water in hotels and remote villages. Again, it has to be clear water to begin with and the SteriPen only kills bacteria and viruses and does nothing to remove chemical contaminants.

SODIS Method

This method for water purification is something that would be useful in an urban environment where chemical contamination is not an issue. SODIS was invented by a Swiss humanitarian group and it is now used throughout third-world countries to provide safe drinking water for thousands of people.

The method involves filling a clear plastic water bottle and then placing it in the sun for six hours. This allows the UV rays to kill the bacteria, viruses, and critters.

Very low-tech and simple and it will even work on a semi-cloudy day. Some things to remember: it works with only with PET plastic bottles and clear glass bottles; water on the cap and bottle threads won't be purified and a straw is recommended to extract the water safely from the bottle when drinking. Like the SteriPen method, you also need clear water to begin with otherwise the turbidity will prevent the UV rays from penetrating.

Still, this would be a good method to file away for an urban setting as one and two liter bottles are plentiful. SODIS has their research and methods detailed on their website including information on how it can be used even in cloud-covered cities like Seattle. For further information, check out www.sodis.ch.

Mechanical

You can use a modern (hand pump) filter like an MSR or Katadyn to treat water and remove any bacteria. These work best if you pre-filter the water through a coffee filter or bandanna. I like the MSR and use it on personal wilderness trips. The advantage with the MSR is that you can strip the unit apart and clean the (ceramic) filter. Most of the other models have pleated filters that must be replaced frequently. Some mechanical filters will also remove chemical contaminants.

Another excellent water filter system that has been used for years in the self-reliance and homesteading community is made by Berkey. These are not intended as portable units for a Bail-Out Kit but for the home. The Berkey filters

come in six models and range in size from 1.5-gallon to 6-gallon models and the internal filters can be cleaned. Just keep in mind that these are gravity-fed systems so it's not like turning on the tap but, hey, in a grid-down situation you will become time-rich anyway.

Chemical

There are three chemical treatment methods available: 1. Bleach 2. Iodine, in either tablet, crystal, or tincture form 3. Chlorine Dioxide

Keep in mind that chemical treatment is similar to boiling water. It is only going to kill viruses and bacteria and does nothing to neutralize other chemical toxins that may be present in an urban disaster when fuel, oil, or other hazardous substances leach into the water table.

Bleach is my preferred method for an urban setting as it's cheap, easy to use, kids can handle the taste compared to Iodine, and it's good for household sanitation (more on this later). I have several friends who have hiked the Pacific Crest Trail over a five-month period while using bleach for water purification on a daily basis.

On our fieldcourses, we use six drops of plain bleach per quart of water. Make sure you use plain bleach as the scented variety has a detergent additive that will make you sick. One of my students, who is a chemist, recommends adding a few drops of food coloring into the bleach solution, so you can monitor its spread throughout the water being treated.

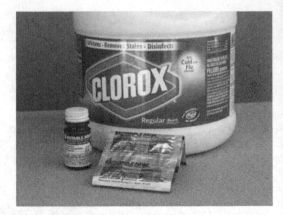

Iodine has been used for years by the military as well as the backpacking community. It can have a gag factor. The key with iodine treatment is to follow the manufacturer's directions as company specs vary. I use the Potable Aqua brand which has 50 tablets per bottle and a shelf life of one year after it is opened. Two tablets per quart of water is the recommended treatment. The only drawbacks: iodine is not good for pregnant women or those with thyroid problems and is not effective in killing cryptosporidium. You also won't get any kid to down water treated with iodine, which is why I don't recommend it for a family survival kit.

A more palatable alternative to Iodine is Potable Aqua's Chlorine Dioxide Tablets. This will give the water a taste similar to bleach (or a swimming pool -and who doesn't like the taste of a swimming pool…). The treatment time is longer with these tablets but the taste is better than Iodine.

Try out a few different brands at home to sample the taste and to make sure there are no side effects before you and your family rely on them in a crisis.

Out of these four methods, I would recommend purchasing a water filter like the MSR which will handle chemical contaminants and then getting a couple bottles of bleach for handling bacteria and viruses. Along with boiling water on your stove (assuming there's power), you will have three means of purifying water and staying hydrated. As we've found on our desert survival courses, if you don't stay hydrated you will become like beef jerky.

Electrolyte Replacement

If you are consuming copious amounts of water each day in an urban survival situation then you will have to add in electrolyte replacement at some point. At home, you can make a simple rehydration drink using two common

household items: In 1 liter of water, mix 1/2 a teaspoon of salt and eight teaspoons of sugar.

There are also some excellent commercial powders you can purchase such as GU20, Vitalyte, and Nuun to name a few. Salty nuts, pretzels and chips can help but downing a bag of Fritos every day is going to reduce the sand in your hourglass.

Replenishing lost electrolytes is essential in hot-weather environments like those found in the Southwestern and Southeastern U.S. where people are chugging water like a lost camel. Failure to do so can result in a dangerous medical condition called Hyponatremia or Water Poisoning, which can be just as deadly as heat-stroke. Hyponatremia is preventable by accompanying voluminous water intake with electrolyte replacement so your sodium and potassium levels in your bloodstream don't become diluted. When I am in the intense heat, I will go through 2–3 electrolyte packets a day. The rest of my sodium/potassium intake comes from my chow.

Alternate Water Sources in Your Home

If you foresee a disaster hitting your region and know the power could be disrupted then fill up the bathtub. This will provide you with gallons of water in an emergency.

A hot water heater can contain 30-40 gallons of water. The tank should be turned off first and then the water tap at the bottom can be slowly opened. Remember this is a HOT water heater so wear gloves and be careful. If you've never cleaned your tank, then it may have some

silt and dirt in it so strain this out and then use a water filter to purify it.

Urban survivors I've interviewed over the years have also mentioned getting water from their toilet tank (not the bowl….). It's only a few gallons but is one more resource to consider.

———— Other sources people have used under extreme conditions after purifying- swimming pools, aquariums, lakes, rivers, rainwater, and their office's hot-water heater. If you have read my other books on survival, then you know that solar stills, tree stills, drinking urine and eating cactus have no place in the real world.

Snowmelting Devices

On overnight winter treks, I always bring a pillowcase that I turn into a snowmelting device later at camp. Simply pack the pillowcase with snow and hang it off a branch near (not over) the campfire. A pot below can catch the dripping water. It normally takes about 30 minutes to fill a quart of water this way so we keep the device going during our evening fire. One woodsman from Michigan I know, prefers using a mosquito head net instead of a pillowcase as his snowmelting device. Just be cautious not to get the head net, which is made of nylon, too close to the fire or it will melt.

Other methods that I have used include the "snow marshmallow" where you take a large, football-sized lump of snow, and place it on a stick anchored near the fire. You can also use a few (hopefully clean) socks or a bandanna stuffed with snow and hung by the fire. I have also heard of survivors using black trash bags and reflective emergency blankets with snow on top to passively melt snow in the sun.

If there is a lake nearby that is safely frozen over (2+ feet thick), then we have chopped holes through these using an ax and retrieved water. The next day the hole has to be chopped back open again.

On short day hikes, I bring along a Nalgene water bottle that is covered from top to bottom with black duct tape to provide me with a passive solar snow-melting device. I have another (lightweight "whiskey" flask) bottle that has a 3' loop of webbing taped on so I can wear it around my neck and inside my parka where my body heat converts the snow into water while I hike.

If you are going to melt snow in your cooking pot over the fire or camp stove, be sure to add a little water first unless you like the taste of burnt snow.

Making a "Water Map" for Your Region

While talking with a former survival student of mine about disasters, the issue of water resources came up. Concerns like: what to do when your tap runs dry during or after a disaster and where are the best sources outside of your home, outside of your city, and which lakes/rivers/springs are nearby?

Every region of the country has its survival concerns to factor in with regards to disaster preparedness, and here in the Southwest where I live, water figures prominently on my list as you might imagine.

I have a "water map" that I've made up over the years for my region. It is a Forest Service map that has red-ink marks for all the water sources I am personally familiar with and also handwritten notes on their reliability, access (only in summer?), and purity. This is something I think everyone should do for where they live, if water is a major survival concern (maybe not you folks in the Pacific NW or Great Lakes!).

Get a local map, start with a ten mile radius and move it out from there as time permits. Then drive/hike around one weekend and check them out. Are they safe to drink from? What's upstream (farms with pesticide runoff or an old mine)? Is the water year-round or just seasonal? And so on.... Check on your

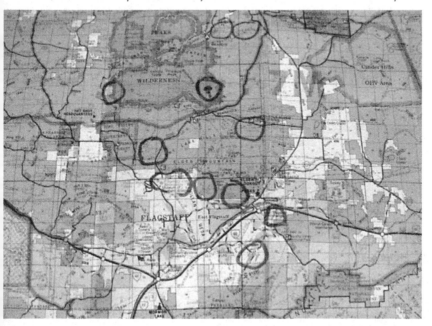

water sources a few times a year or more, especially the primary ones you'd consider using in an emergency.

Now, I'll qualify all this by saying that I live in a high-desert region so my water map and concerns look different than someone in a city. The principles are going to be the same: where, outside of your neighborhood or metropolis, is your nearest reliable water source(s)? For those in a large metropolis you might want to familiarize yourself with city parks and swimming pools. Not that you're going to be chugging out of your neighbor's in-ground pool when the lights go out for an afternoon. This is for long-term considerations where the water tap has been dry for two weeks and your stored water is running low (such water sources by the way would need to be filtered to remove chemical contaminants).

Don't rely on topographic maps to be up-to-date either. Most are dated 1965 or 1983 and out West water availability in the wilds changes each season. I remember doing a land navigation exercise with my students where we trekked to a designated "water tank" on the map. When we arrived, we found a rusty cattle trough turned on its side and riddled with bullet holes. So much for slaking our thirst based on the topo map features which, if we had bothered to check, was dated 1980!

And yes, being prepared and having quantities of water on hand at home are essential but there's no substitute for "local knowledge" of nearby water sources for a potential long-term situation. Such knowledge costs little and can go a long way if your well-prepared stocks at home run low or are comprised in some way.

Summary

Starter Kit for Water Storage and Purification

- 2, 7-gallon water cubes
- 12, 1-gallon size juice bottles
- MSR Water Filter
- 1 gallon of plain bleach
- 12 Electrolyte replacement packets per family member
- Create a "water map" for your region

Food Recommendations and Storage

How Much Food Should I Have?

There isn't a cookie-cutter answer here. Again, it all depends on your budget, family size and space. I have one student who lives in an apartment in downtown Chicago who has very little space but she has managed to squirrel away a 14-day supply of food and water for herself. A college student I know, living in a cramped dorm, only has room for four days of supplies while some Mormon friends of mine have a ranch with a two year supply for both people and livestock. Again, it's dependent on your situation. What I recommend to my clients is that, space permitting, start in the area of having a 30-day to three-month supply of both food and water and then inch it up when your budget and living conditions permit.

Types of Food to Stock Up On

I sometimes have people in my survival courses who confide that their emergency food storage consists of six months of MREs (Meals Ready to Eat) or a 30-day supply of Ramen Noodles.

There's nothing wrong with having the above chow on hand in limited quantity. I carry a few MREs in my truck for a roadside emergency but I sure don't want to live off such synthetic stuff for a couple of months. I also know that the color will drain out of my kids' faces if they are ever confronted with a MRE again. Save the MREs for short-term situations or your Bail-Out Bag or the emergency kit in your office and invest in foods that you normally eat at home.

Following, is the approach that I have used on our long-term

survival courses where we are out in the field for a few months, and occasionally supplementing our diet with any wild caught foods or foraged plants. These staples are what we also utilize at home and what I recommend to my students for prepping their home pantry.

There are three food types to plan for:

Dried Goods

Rice, oats, lentils, millet, beans, green peas, brown sugar, coffee, tea, pasta, flour, dried soup mixes, wheat and barley are some of what we have in storage. Personally, I have lived on rice and lentils for dinner and oats for breakfast for up to 21 days straight on long-term wilderness treks and my body did extremely well. Rice and lentils provide a complete protein and plenty of carbs. Tabasco was a good friend with such an unchanging menu.

The general rule is that we go through one cup, per person, per meal at the low-end and a cup and half for those with higher than normal caloric intake or due to cold-weather. On fieldcourses where we have all guys, then the food consumption goes up by about 1/3 compared with a women's course. Something else to think about if your family makeup is either boy or girl dominant.

Canned Goods

We use this as one of our main staples at home and not so much on our fieldcourses due to weight issues and the cans freezing in cold weather.

Stock up on your favorites by purchasing three cans more of what you normally eat on each trip to the grocery store. In our household, we have the following on hand: corn, pinto and black beans, green beans, pumpkin (great for mixing into pancakes and muffins), pineapple, tuna, diced tomatoes, ham, chicken, tuna, pears, and squash.

A friend and fellow survival instructor, Mike Masek, introduced me to this fantastic-tasting, low-tech dinner that uses canned goods and one that we now have as a staple on our fieldcourses as well as at home once a week. Big hit with the kids too! If we can toss in some meat or chicken then that's a nice bonus but not necessary.

....

Masek Southwestern Stew (recipe good for 8 people)
  8 cans of corn
  8 cans of pinto or black beans
  4 cans of diced tomatoes (salsa will work too)
  Big handful of fresh cilantro, basil, or similar dried herbs
  1 cup of water

  Heat for 10 minutes until warm. Can be served cold as well in the event you don't have a stove.

....

Big or Small Cans?

For large groups of 8-10 people we use the 64 ounce cans of beans, corn, etc...but we find a lot of food gets wasted with numbers smaller than this. Assume you won't be able to refrigerate large quantities of leftovers in a grid-down situation so purchase the smaller cans and then rotate your stock each month. The FiFo brand can-tracker, which can hold up to 54 cans, is a good tool for ensuring your items get rotated.

Dehydrated and Freeze-Dried Food

My advice is to go to an outdoor gear shop and purchase a few different brands of dehydrated and freeze-dried meals first before investing a few hundred bucks in supplies that may have a considerable gag factor. These meals are also intended for people who exert themselves (backpacking, canoeing, etc…) and need to replace lost salts so some of the meals contain up to 56% sodium. Not recommended for life at home, long-term.

Mountain House and AlpineAire are good brands with many fine selections that can be found in outdoor gear shops. My experience with freeze-dried eggs and breakfast entrees has forever turned me off but I've found the dehydrated eggs sold in bulk at health-food stores to be very tasty.

Walton Feed sells high-quality foods such as dehydrated veggies and even canned butter. We have used their products in our meals at home and it tastes like food from the grocery store. I'd highly recommend the dehydrated foods or you can buy an Excalibur Food Dehydrator and make your own. I have a three-tray Excalibur for making jerky on a weekly basis and it's the best commercial dehydrator on the market.

Most of these brands have a shelf-life of five years and require a cup or two of boiling water to rehydrate. I will take these any day over MREs but I know folks who have their favorite MREs too. Again sample things and see what tastes good, especially if you're planning on living on these for a few months.

Putting It All Together: A Sample Menu For The Week

We always alternate our meals so one day is dried goods consisting of oats/fruit for breakfast and rice/beans for dinner. The next day we have pancakes with (canned) pumpkin or pineapple mixed in for breakfast and then the Masek Southwest Stew with canned goods for dinner. Day three is oats again for breakfast and pasta or stir-fry (using dehydrated meat and canned veggies) for dinner.

Frankly, I eat oatmeal at almost every breakfast when I'm at home or in the field with some fruit tossed in. You already know what works for you so stock up on those items on each trip to the grocery store if the budget permits.

What About Lunch?

In long-term survival, you should reserve the propane, fuel, or firewood for breakfast and dinner. Lunch can consist mainly of dry goods such as jerky, crackers with peanut butter, tuna, canned ham, dehydrated apples or carrots, leftovers, and cheese (as a side note, cheese lasts for ten days in the field without refrigeration. Just buy mild cheddar as it'll be sharp cheddar by the end!).

Special Considerations

With our Just-In-Time economy, you don't want to be at the mercy of the resupply trucks, especially if a blizzard has shut down your city for a week or a long-term disaster affects your region for the next month. If you are a parent then have a 30-day supply of formula, diapers, meds, and extra baby-food on hand.

Vitamins

These are an essential part of keeping your body balanced so don't forget to stock up on a 30-day supply of vitamins.

Spices

Did I mention Tabasco already? Talk about a lifesaver and a great barter item! Aside from that, having the usual players on hand like salt and pepper, will be a big help. I also relish dried basil, cilantro, and onion flakes so we have enough for taking care of plenty of evening meals.

Home-made Jerky

Other than using a food dehydrator, which requires electricity, here are a few methods we use on a regular basis when out in the field and at home. Meat preservation is a skill area that gets into what I call the realm of Bushcraft. These are the long-term skills of living with the land that go beyond mere survival. It all depends on where you live and what the humidity level is whether you smoke or jerk meat.

Where I grew up in the Great Lakes, we had to use the pioneer method and construct a primitive smoke-shack that would smoke and dry out the meat over a 3-4 day period. This was necessary because the high humidity made air-drying on an open rack next to impossible.

Where I live now in the arid Southwest, making jerky is caveman simple. We cut the meat (deer, squirrel, beef from the grocery store) into 1/8" or thinner strips, douse with some spices or hot sauce, and then place on a drying rack in the sun for 8-10 hours and voila- the best jerky you'll ever taste. My favorite "marinade" for jerky is to soak the batch in a mix of barbecue sauce and honey before hanging it on the drying racks. If I am going all primitive, then the meat just goes on the rack minus the marinade. Fat must be trimmed off so it doesn't later go rancid.

I once used this method with a group on a desert survival course in May and our strips of meat were dry and brittle within six hours! The test to see if your jerky is done is whether you can crisply break a piece in half.

You will find that indigenous cultures throughout the world all trim their strips 1/8" thickness or less as this prevents flies from laying their eggs in the meat and also speeds up drying.

I'd say look into the archeological record for your region and see what methods were employed by the natives. It is either going to be a smokeshack setup of some kind or air-drying on open racks. Practice with some cheap cuts of lean beef from the supermarket and see what works in your neck of the woods.

Deprivation Issues

We've had students on fieldcourses over the years that go cold-turkey for the week on coffee, smoking, chocolate, their favorite food, etc....They are usually a train-wreck for the next three days. It manifests itself in the form of irritability, lack of sleep, and being mentally distracted. It can turn the group dynamics into a horror show. This usually levels off my mid-week but in the meantime, I'm scratching my head wondering if I missed something on their health form.

One guy, from Europe, was used to drinking five cups of espresso every day back home and had been doing so for 20 years! My instructors and I figured he'd have a heart-attack if he was deprived of caffeine for even a day. So severe was his addiction that he actually snuck in some instant coffee packets for the solo phase of the course and would wedge one between his lower lip and gums like tobacco.

My point here is, in addition to the outward stress of a catastrophe, there may be internal stressors from deprivation/addiction going on in your circle of fellow survivors that may wreak havoc on their mental attitude and ultimately the group's cohesion. For some people, especially the younger generation, being without their Blackberry, iPod, or Wii might also be enough to tip the scales.

Keeping them busy by improving their surroundings, working manually with their hands, and staying on top of hydration are a few of the things that will help with a smoother transition for most people.

Pets

Think you're going to be turning Peanut and Scraps loose to fend for themselves in a crisis? Not likely, unless they were raised on a ranch. Two of my dogs are domesticated crème-puffs that were reared in an animal shelter and the other two are from the Navajo Reservation where they lived like coyotes for the first part of their lives. The latter are pretty exceptional hunters when we are on the trail. However, I am not going to let them run wild in town if the feed stores empty as they're likely to get shot or run over. Plus, they are cherished members of our family.

Preparation for yourself also involves preparation for the other family members such as your dogs, cats, goldfish, iguana, etc... Same rules apply—stock up on an extra bag or two of chow when you're at the pet store next time. Monitor how long it takes your dog to consume one 40 pound bag and then get enough to last at least a month. Then if you bag a squirrel on the fence-line, your trusted canine can share in the remains but at least he will have that bowl of crunchy vittels to fall back on.

If the Mountain Men Could Do It....
Many students come to my survival courses thinking that living off the land can be likened to walking with a cart down the produce section of the grocery store picking out what they want for dinner. I wish it were so but in reality there's a reason it took a tribe to feed a tribe and why hunting

and gathering was something our ancestors did from birth to death. The quest for food in the wilderness is a demanding study. I have found that some days can bring a bounty while others an empty stomach.

If you are versed in hunting, fishing, foraging, or trapping then you will fare much better in a grid-down situation but you will still want to have a stash of food on hand at home for your primary sustenance. The time to learn how to live off the land is NOW, not when the grocery shelves empty during a pandemic or disaster. Such skills are best learned from one versed in them. Seek out experienced hunters/fishermen amongst your family or friends and spend time learning the basics. Take a class on how to safely identify edible plants from your local nature center or arboretum and then incorporate three common edible plants from your region into your diet each summer.

If it's been a while, dust off the old fishing pole and go out one Saturday morning to reacquaint yourself with how it's done. By far, the best skill that you can learn, and one that is far more reliable in its yield, is trapping. Not for getting a mink coat but for putting meat on the table when all the big game is depleted. Seek out a trapper in your region and spend time with them learning the craft. It will be time well spent. The trappers I know have a tremendous land ethic and know wildlife like few people today.

Lastly, my survival school and many others, offer courses in how to live off the land using modern and traditional methods. At my school, we offer several immersion courses that cover the basic tools and techniques for procuring wild game and utilizing edible plants. Check out the appendix for a listing of other survival schools around North America.

Summary

Food starter kit for 1 person for 30 days:

8 pounds brown rice
8 pounds lentils
4 pounds pinto beans
5 pounds of noodles/pasta
4 jars of spaghetti sauce
2 pounds of salt
8 pounds of brown sugar
10 pounds flour (self-rising)
4 pounds of your favorite dried fruit- raisins, apples, peaches, etc…
2 pounds old-fashioned oatmeal
4 jars of strawberry or grape jam
8 jars of peanut butter
8 boxes of whole-wheat crackers
30 multi-vitamins
30 Vitamin C tablets
2 large bottles of Tabasco
1 pound of salt
2 ounces of pepper
2 ounces garlic powder
2 ounces onion powder
4 boxes of powdered eggs
4 boxes of powdered milk
1 pound cooking oil
15 cans of tuna or chicken
15 cans of ham
Plenty of your favorite sweets- chocolate, cookies, etc…

This list covers just the basics so add in your personal favorites with regards to grains, cereal foods, spices, and meat. Don't forget an old-fashioned (non-electric) can-opener.

A FAMILY FIRST AID KIT YOU CAN LIVE WITH

Having field-tested first-aid kits over the years on extended wilderness trips and being a parent who has dealt with my share of late-night illnesses, I recommend purchasing a quality wilderness first-aid kit for your urban medical needs. These kits are far superior to the generic, band-aid heavy kits found at the local pharmacy. They were designed with remote medicine in mind which is what you will have in an urban disaster since you will become your family's first responder if the 911 system is ineffectual.

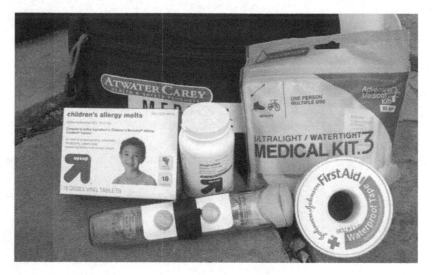

As mentioned earlier, I am a big believer in layering your survival gear and the kit that follows is for the home while we have another in our vehicles and a streamlined version in our Bail-Out Bags.

There are two companies that make outstanding first-aid kits in both individual and group sizes. They are Atwater Carey and Adventure Medical Kits. For my family, we have the five-person Atwater-Carey Kit and I have augmented it with the items listed below. Not all of the items fit into the kit so we have an ammo box that fits everything. In our personal family kit we have a blend of modern and traditional medicines.

Lastly, I can't recommend enough taking a wilderness first-aid class and obtaining hands-on skills in remote medicine. The Wilderness Medicine Institute offers 2, 4 and 9-day classes throughout the world on a monthly basis and there is probably one offered in a city near where you live.

Immodium – for diarrhea, 24 tablets
Ibuprofen, Aleve, or similar painkiller –100 tablets minimum
Benadryl – critical for bug bites & severe allergic reactions. Get the Fast-Melt
    kids version, 48 tablets
Aspirin – for the usual aches and pains. Can also be used for dogs, 100 tablets
Arnica – herbal salve for sprains, bruises and muscle strains
Tweezers – get a quality pair with thin, flat-nosed tips for removing splinters
Duct tape – for instant band-aids, covering blisters, fixing gear, and a hundred
    other uses
Sani-Wipes or Hand-Sanitizer
Polysporin antibiotic ointment, two tubes
Fingertip and Knuckle Bandages – these hold up longer, 48 bandages
ACE Wrap – hard to find stretchy fabric when you need it
PriMed Gauze – simply the best gauze material on the market for dealing with
    intense bleeding
Triangular bandage – myriad uses for tourniquets, slings, head wraps, straining
    water
Steri Strips – for closing wounds until you can get stitched up by a doctor, 48
    Strips
Caromeds Bleeder Pack – for traumatic wounds
Irrigation syringe – a must have item for first-aid kits. Great for blasting the
    nasty germs out of wounds
Temparin Dental Repair Kit for patching up a lost filling (though I know of a
    few people who have used candle wax for temporary fillings until they could
    get to a dentist)
Oil of Clove – for dampening the pain associated with lost dental fillings
Tea Tree Oil – has anti-fungal and antiseptic qualities for use topically on cuts/
    infections
Hops herbal tincture – for use as a sleep aid when sick
Emergen-C – used in electrolyte replacement and vitamin intake
Children's Motrin or Ibuprofen – if you have kids

Children's cough medicine and sore-throat lozenges
Daily Multi-vitamins
Bag Balm – the best there is for dry, cracked skin
Where There Is No Doctor Book

Personal Medications to Stock Up On

During a short-term emergency lasting a few days, the pharmacies may remain open but in a long-term disaster these will quickly be depleted or looted. Is someone in your family diabetic or on blood-pressure medicine? Does your child have a history of asthma and need inhalers? Are you dependent on contact lenses or prescription glasses? Stock up <u>now</u> on personal prescriptions that are critical to your health, ideally striving for a one-month supply beyond what you normally need.

Lastly, who in your neighborhood is a nurse, EMT, former combat-medic, or physician that you could request help from during a family medical crisis? Chances are they may be called away in a short-term disaster but long-term they may be at home with their families.

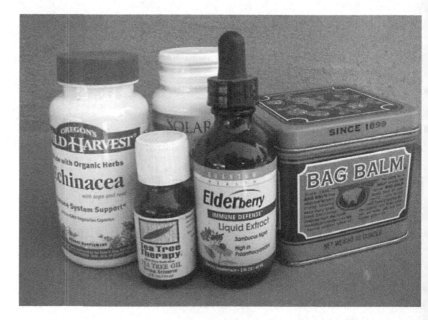

Medicinal Herbs

I am not talking about California-style medicinal herbs here. Being involved in the study of edible plants over the years has led me into the world of herbal remedies. These are time-tested herbs that sustained our ancestors long before the advent of modern antibiotics and synthetic medicine. I have incorporated many herbal remedies into my family's first-aid practices and over the years we have used common household herbs such as ginger for stomach aches, cayenne pepper as an anti-coagulant and antiseptic on cuts, and licorice root for sore throats.

Learning to make your own herbal tinctures, salves, and flu remedies can be a rewarding way to boost your self-reliance skills and provide a greater connection to the land even your backyard. You would be surprised at how many herbs and edible plants can be found within a short walk of where you live, even in a large city. If you want to get started, take a class with an experienced teacher who can show you how to safely collect and process herbs. There are many schools that provide weekend training courses in the basics and even sell herbal first-aid kits that you can use to augment your family's kit at home. For some people, they are content to know just a dozen common herbs for their region, and for others, herbal medicine becomes a lifetime study.

Incorporating both first-aid skills and traditional herbal medicine gets you dialed into both worlds and provides you with more tools to employ during a crisis. Plus, you don't need a prescription for gathering and making your own herbal meds!

Recognizing Hypothermia

The classic identification method for cold-related injuries is to look for the "umbles." This is a term that comes from the wilderness medicine community and refers to a change in a person's level of consciousness and physical coordination. In essence, if you see someone stumbling, mumbling, bumbling, and fumbling, then they may be experiencing hypothermia. Conversely, the umbles are also associated with heat-related injuries such as heat-exhaustion and hyponatremia.

Keep in mind that most cases of hypothermia happen in 50 degree F weather and not just in the depths of winter. Many people who succumb to it are dressed improperly wearing cotton clothes and there more fatalities from hypothermia in urban areas than in the wilderness. Wet cotton will contribute to hypothermia as it wicks your body heat away from you and fails to insulate unlike wool and fleece garments. So, avoid the jeans, sweatshirt and other 100% cotton fabrics.

If you are experiencing hypothermia or notice someone in your group with the umbles do something about it NOW and prevent it from getting worse. Light up the wood stove, get out of the wind, change into dry clothes/footwear, or get into your sleeping bag. Eliminate the conditions that are creating the problem. Then stoke your internal furnace with some high-calorie, high-fat foods such as peanut butter, chocolate or soup with cheese.

My usual cold-weather recipe for when I start to get chilled is to have a cup of hot cocoa with a tablespoon of butter in it. I carry a thermos with this brew on my winter day hikes. The combination of fat and sugar helps to amp up the metabolism.

Hypothermia is the number-one killer of people in the outdoors the world over, so dress properly to deal with wet, windy conditions, bring high-energy foods, stay hydrated, and keep an eye out for the umbles.

Off-Grid Medical Books

One area that is often overlooked in the study of survival is human physiology and remote medicine. Having the right gear, a quality blade, proper clothing, and training are essential but too often preppers forget to verse themselves in the physiological aspects of something as simple as hypothermia as well as other cold-related injuries and hyperthermia (heat-related injuries).

There is sometimes too much emphasis in magazines and survival forums about the best headlamp or the latest tactical blade but little discussion on what it takes to sustain the human engine in extreme temps in the wilds or even under long-term conditions associated with a civil breakdown.

If you don't have the time to take a course in wilderness medicine then, at the very least, obtain a copy of the books below and digest the material on heat & cold injuries.

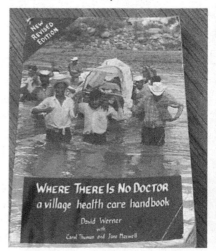

WHERE THERE IS NO DOCTOR
a village health care handbook

David Werner
with
Carol Thuman and Jane Maxwell

The top three books I use are:

Medicine for Mountaineering—An excellent manual on everything that can befall you in remote settings.

Where There Is No Doctor—a must have book for both urban and wilderness skills. This book was developed for Peace Corps workers living in remote regions. As a parent, I highly recommend having it at home for an emergency and in the Bail-Out Bag. This book was very helpful during those late nights when my kids were sick.

Survival At Sea—if you want to understand the science and physiology behind what makes us tick under extreme conditions, this is the book. I reference this one often and use the data in my courses when teaching about heat & cold injuries.

Remember, survival is as much about understanding your own physiology as it is about the gear in your pack. Train the head as well as the hands.

Mental Health During Long-Term Disasters

From being cooped up with groups on fieldcourses in cabins, caves and snow shelters for days on end during winter storms and foul weather, here are a few things I have learned that can help with maintaining daily Positive Mental Attitude (PMA) for yourself and a group.

• Staying clean—have a daily ritual of washing face, hands and privates.
• Having a routine (this is what POWs have said got them through each day): breakfast, cleanup, reading time, manual labor, lunch, rest, writing, manual labor, dinner, storytelling/handcrafts, sleep
• Sleep and catnaps
• Story swapping (yes, even for adults) a few times a week in the evening
• Playing cards or board games—have two game nights a week
• Carving or crafting something—this allows you to take control of your surroundings
• Having a detailed plan/common goal you are working towards

Lastly, I give a lecture in my wilderness survival classes about survival psychology, in particular, what I call the *One Thing*. This is the force that drives your life. Are you a parent, a caregiver, or have someone for whom you would walk through a hail of flying daggers for? This *One Thing* is what motivates people during times of duress and extreme survival. It is what galvanizes your will and causes you to shout out— "I will be here tomorrow when the sun rises and the next morning after that, no matter how ugly things get today!"

You will have highs and lows with PMA. We all do. It's a part of the human experience. When you have the lows, however, is when you must do something physically to take control of your surroundings and regain that sense of purpose. Gather some more firewood, spruce up your sleeping area, or simply washing your face and hands can turn your attention outward away from the negative juju inside getting you down. Then, swig down a can of Toughen-Up and remember what that *One Thing* is in your life that will pull you through.

Summary
First-Aid & Mental Health Starter Kit

Two-day Wilderness First-Aid class
A wilderness first-aid kit tailored to fit your needs and family size
Copy of *Where There Is No Doctor*
30-day supply of prescription meds
2 pair extra prescription glasses or back up contact lenses + cleaning solution
Board games, journals, coloring books for kids, and playing cards

HOME SECURITY AND PERSONAL DEFENSE

As with all other areas of survival preparation, the layering method is applied to the security of both your home and the immediate area outside of it. Home defense is about more than having a shotgun. Much of the information that follows comes from time spent with personal security experts, combatives and firearms instructors, and interviews with law enforcement personnel.

First off, keeping your preps low-profile is the key to initially preventing a lot of problems. If you are known as the "survival guy" or the stockpiler, then guess where folks are going to come knocking on day four of a catastrophe when their stomachs are rumbling?

Security Measures for Outside the Home

Think like a thief when examining your home. Are there bushes or low-walls to hide behind? Which part of the house is most obscured by darkness? Is there a fence to contend with? Any cat or doggy doors to gain access? Are there alarm company or dog signs present? Do the doorways have exterior lighting? How close are your neighbors' homes and are they actively living there?

Visual Security

Solar-powered lights and walkway lanterns, especially those that are motion-activated are best for augmenting your existing electrical lighting. Pay

attention when pulling into the driveway if your exterior lights aren't working.

When walking out of the house or your vehicle at night, carry a high-beam tactical flashlight such as a Fenix or Streamlight brand as these can provide momentary blindness in your attacker. Ever had a policeman shining his flashlight in your face at night? They do that to gain the tactical advantage.

Having the heightened awareness of a puma is critical when walking from your house to the car or vice-versa as most attacks will occur in this zone based upon crime statistics. Avoid leaving shovels, rakes, and other implements outside your house to reduce weapons of opportunity for criminals.

External Doors

You can purchase metal security doors at big-box stores that can be painted to look like a normal door. The next step to further fortify the door is to use 3" screws in the door jam attached to the stud. There are also wireless

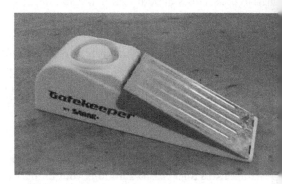

door stop alarms that are battery-operated and wedge behind the door. I use these in hotel rooms when traveling. GE makes a number of motion-activated, battery-powered alarms that mount on the inside of a door or window which can serve as a low-cost alternative to a pricey, home-alarm system.

Securing Windows

There are several companies that make security window film that is designed to keep a window intact after it has been shattered by hurricane force winds or by smash-and-grab robbers. It comes in 8 mil and 12 mil sheets that you can apply yourself. ShatterGuard and 3M are two companies to research.

Another approach is to fit the interior portion of windows with security bars. One company, Mr. Goodbar, provides excellent security bars with an ornamental appearance (vs. prison appearance) and are easy to install from inside the house.

The low-tech method of placing a 2" x 2" piece of wood in the window and patio door track will help prevent forced openings. As noted with doors, there are battery-powered window "squeaker" alarms available as well.

Interior Security Measures

I will break down each one of the following strategies but here is how the layering approach to your home's interior security could look: having a dog;

then having a firearm on you or in an easily reached location; then having pepper spray in each room; then having a safe-room with a solid-core door and locking deadbolt; then having a security-pack in the safe room.

Sound paranoid? Remember that one of anything critical is a poor system. This is why professional soldiers and law enforcement officers have so many back-up measures and gear in place to handle Murphy's Law. A layered approach will allow you time to get to your safe room, exit the house, or engage with your weapon.

Dogs

I'm not going to do a rundown of AKC breeds here but suffice it to say that a dog is a tremendous asset, especially if you live alone. I've always preferred the herding breeds such as Australian shepherds and border collies. Labs, German Shepherds and Rottweilers are also outstanding. A dog is an awareness multiplier and will allow your senses to extend throughout your house while you sleep, in addition to being a good deterrent for prowlers looking for an easy smash-and-grab house.

Firearms

Many authorities have written about the type of firearms to have at home for self-defense. Owning a firearm, whether a pistol or rifle, brings with it tremendous responsibility and the need for training. Otherwise, you will be a danger to your family, yourself, and your neighbors.

There are endless websites and forums focusing on which pistol or rifle are the best. Start by getting training in firearm laws, safety, shooting, and handling skills. The NRA offers classes nationwide that will cover the foundation skills and legal

issues. You may also have someone in your family or circle of friends who is former military and can show you the basics but look into your state and city laws.

After this, I would take your training a level further and seek out an instructor versed in tactical firearms training who can show you what's involved in the dynamics of an actual gunfight and how to shoot while moving, from your car, in the house, etc.... One of the best training schools I know of is Suarez International and they have instructors throughout the U.S. and internationally.

Regarding what type of firearms, it's hard to beat the baseline three: pistol, shotgun, and .22 rifle. Again this is a place to start but it will provide you with both defensive and food procurement capabilities.

Pistol

If you only go shooting once a year then a revolver is pretty fail safe compared to a semi-automatic which requires you to know malfunction drills and have greater familiarity with its mechanics. For revolvers, a 4" barrel .38 or .357 (or combo) are common calibers and the ammo is easy to obtain. For a semi-auto pistol, Springfield XD, Sig Sauer, Glock, and Ruger are all outstanding and battle-proven.

I'm sure to get a bunch of emails with exclamation points attached, such is the caliber controversy, but I recommend going with a 9mm which is more affordable than the higher calibers and has been proven to be just as effective by operators in real-world conflicts. Purchase a dozen or more magazines but check into your state's laws with regards to magazine capacity. Pick up 2000 rounds of ammo with a mix of defensive (hollowpoint) and target (full metaljacket) rounds.

Shotgun

A shotgun doubles as home defense and large/small game hunting. The pump-action Mossberg 500 and Remington 870 are two of the more reasonably priced models available and both have a long history of reliability. Get a Kick-Eez recoil pad and your shoulder will be thankful. A sling would be another essential accessory along with 500 rounds of assorted buckshot and 100 rounds of slugs.

.22 Rifle

This is the survivor's best friend when it comes to target practice and small-game hunting. In addition, ammo is cheap and lightweight. I've taken more small game over the years with a .22 than with anything else. Outfit it with a scope and you have an excellent tool for feeding yourself in the wilds. I prefer the Marlin Papoose which is a takedown rifle or the Ruger 10/22. The latter rifle is probably the most talked about .22 on forums and has been around a long time. Pick up a sling, 12 spare magazines and 2000 rounds of ammo for starters.

Holsters

Regarding carrying your sidearm, I have some holsters that are for concealment and some that are for wearing in the open when I'm in the backcountry.

I've probably gone through twenty holsters over the years trying to get that perfect match for my needs. Make sure it's a comfortable fit as nothing is more unpleasant than a cheapo holster straining the back and putting a dent in your side.

Cleaning Supplies for Firearms

But I saw John Wayne clean his gun with dirt... You make do under extreme conditions (or in a rigged, action-flick) but in real life, you want critical tools to last so stock up on basic cleaning supplies. For starters, purchase four spray cans of Gun Scrubber and four spray cans or bottles of Rem Oil and then get a few Gun-Bore cleaning snakes to match your caliber firearm.

Less Lethal Defensive Weapons

OC Spray

Two other, less lethal, tools for self-defense include OC and pepper spray. Get the spray that is delivered in a stream not a mist as it will allow you to keep greater distance between you and the attacker. Sabre and Fox Labs are two reputable companies with proven track records amongst the Law Enforcement community.

Knife

A knife can be a tremendous deterrent to fend off a potential attacker, requires no ammo reloads, and can be employed where a gun might be too risky for innocent bystanders. The template for blade training has already been worked out over the centuries by the Filipino Martial Arts and Indonesian Silat Arts. Either of these would be a good place to learn how to use a knife for self-protection.

Renowned combatives and blade instructor Tom Sotis teaches seminars around the U.S. in how to prevail during a street attack using a handful of practical skills and not the gymnastic, jump-kick routines from Hollywood. Otherwise, look into combatives training classes in your city as these focus on simple, efficient moves that can be learned in a short period of time and provide you with realistic skills that work under pressure. As with firearms, each state has its own knife-laws so check into these before carrying a blade in your pocket or on a belt sheath.

Multi-Tools

Multi-tools, like knives, are all about personal preference and the nature of the tasks you are undertaking. My primary blade has always been a Swedish Mora but there are times when the various gadgets on a multi-tool are helpful. The one I use the most is the Wenger brand Swiss Army Knife, he Evolution model in particular. This has a handy folding saw along with the usual features (screwdriver heads, awl, tweezers, etc). Mine cost around $25 and s a streamlined model of the older, bulkier versions of Swiss Army Knives.

I also have a Leatherman Wave in my truck's glove box for any vehicle issues. I have found that the most important feature for my line of work is the folding saw that these multi-tools provide. The other features are not as critical (of course, that toothpick feature is nice after a dinner of wild game....). There are so many variations of multi-tools so look at the features you need the most—do you really need scissors or a corkscrew in the wilds?

Also, keep in mind that a multi-tool is <u>not</u> a knife, it is a multi-tool! So, carry a quality fixed-blade in addition to a multi-tool. The fixed-blade will allow you to split firewood, whittle, and handle the heavy-duty chores. Carry at least two blades and remember the old Scandinavian motto: "A knife-less man is a life-less man."

The Safe Room

This is a designated room in your house that all family members know to retreat to in the event of a home invasion or an emergency. Preferably it has one window for egress. This room should have a solid-core door with a lock and 3" screws in the door jam as with the front door features discussed earlier.

In a closet in this room, you stow a fire-extinguisher, flashlight, phone, bottle of drinking water, and pepper spray or firearms (in an easily accessible lockbox to keep it from younger hands).

The Nighttime Security-Pack

Mark, a student of mine, described a break-in at his home a few years ago. He was asleep upstairs with his wife and infant son, when he awoke in the middle of the night to the sound of breaking glass. At first he tried to deny it was happening, thinking that it was a tree branch clanking against the kitchen window. He got up and walked towards the steps to listen when he heard the sound of more glass shattering. His wife dialed 911 and gathered up their child.

Mark welled up inside with anger knowing he had to protect his family but knowing that he didn't know how to do that or have any weapons to use. He started frantically searching for a weapon to use and reached into a nearby closet for his golf clubs. At this point, his frustration and anger caused him to yell down the stairs, "Hey who the Hell is there!" as he turned on the stairs light. At this point, Mark said he heard the burglar drop something and then run off through the backyard. He and his wife remained upstairs until the police arrived about ten minutes later.

They were lucky. Mark told me he and his wife now have a security plan in place, know the individual actions to be taken, have created a safe-room upstairs, received firearms training, own a pistol and gun lockbox, and sleep with a security-pack on the bedpost each night. Living in fear, hardly. Empowering yourself never to be a victim, exactly.

The time to learn to sail a ship is when the sea is calm. Then when the sea is choppy you know better how to navigate. As noted in the above story, you don't want to be scrambling around, half-asleep, trying to find a weapon and flashlight while a potentially violent criminal (or three) are headed your way.

So what's in a security-pack? Most firearms and personal security experts recommend having the following items: pistol, six spare magazines, tactical flashlight, knife, OC spray, cell phone, trauma kit (you might not be the only one sending bullets flying during a violent break-in).

Does this pack look sound like an example of layering? You bet. Maybe the noise in the backyard is from teenage hooligans having fun so you blind them with the flashlight and some colorful language. Maybe the ruckus in the attic is from raccoons so out comes the OC spray. Or you hear two meth-heads looting the garage and wait for the police in your safe room, armed and ready to engage them if they breach the house.

A layered security pack provides you with options whereas simply having a bedside gun can create the situation where "if there's a hammer then everything is a nail."

If you have small children, then the pack gets locked up during the day. This is intended as a night-time, grab-and-go daypack containing gear you have trained with for a physical threat.

Fire Safety

Every home should be equipped with several fire extinguishers. One should be kept in the master bedroom and one on a distant side of the house along with a fire extinguisher at each level (basement and upstairs). In addition, if you have an upstairs, stow a folding escape ladder in a bedroom on that level.

Evacuation Routes Out of Your Neighborhood

Egress routes out of your neighborhood should be worked out beforehand and contingency plans for bailing out should include at least two areas to meet up with other family members. Consider how this plan will look during the winter with three feet of snow on the ground or during the rainy season in the summer. What about walking out on foot? Our family plans are very specific—If we go west, we will meet at Uncle Jim's house or the parking lot of the Holiday Inn on Lucky Lane and so on. This information should be a part of your bail-out kits and be written down and stored there and in the glove box of each vehicle.

Discuss your layered security approach and evacuation plans with each adult family member and have an occasional "fire" drill with younger children so they know what is expected of them during an emergency.

Summary

Starter Kit & Areas for Home Defense
(again check your local & state laws)

Outside lighting that is solar or battery-powered
Tactical flashlight & spare batteries
Pepper spray
Folding knife
Have a designated safe room
Create a security pack to fit your needs and training level
Fire extinguisher
Fire escape ladder (if your home has more than one level)

WHEN THE POWER GOES OUT: HEATING, COOLING, LIGHTING AND COOKING

When the lights go out for an extended period you will be faced with camping out in your home. When I was a kid growing up in Michigan, we had a few power outings each winter when an ice storm rolled in and felled electrical lines. My dad would plant us kids in the family room with our JC Penney sleeping bags, mom would bundle us in sweaters and snow pants, and then they would get the fireplace blazing. Dad would then seal off the other rooms and hallway with visquene and tape. He'd slightly crack open a window to prevent carbon monoxide buildup and then turn on a camping lantern.

These outings were short affairs lasting only 1-3 days but the emphasis wasn't on trying to heat the entire house and resuming our previous lifestyle. It was to heat the body and a very small space. Both of my parents grew up during the first Great Depression and knew well what a life of austerity and improvisation looked like. For us kids, it was an adventure but for the adults it was like stepping back in time to a life of perseverance.

When the power is out long-term, focus on heating (or cooling) the body and not the house. The Japanese live like this full time emphasizing personal warmth over heating a large space. If it's wintertime get out the down jackets, hats, snow pants and sleeping bags or blankets. Gather up family members in one room for sleeping to concentrate heat. Remember to have 24-hour ventilation by keeping a window slightly ajar, especially if a wood stove or propane heater is involved. Seal up the rest of the windows using duct tape/tacks and visquene or a bedsheet.

At night, employ the old camping trick of placing a warm bottle of water at the bottom of each sleeping bag. Then go to sleep with a wool hat on your head and food in the belly which will keep your internal wood stove cranked up.

Propane Heaters

Some ranching friends of mine use a propane heater to warm their 12' x 12' bunkhouse and it works perfectly for this small setting. I've used them in canvas tents for a few hours when we don't want to run the wood stove. These devices are intended for small spaces and won't heat an entire house. The nice thing is that the propane tank can also be used as fuel for a camp stove or lantern. There

are "tree" fixtures you can purchase that will allow for multiple branches off the main propane tank as well.

In mildly cold weather, you can extend the life of your propane fuel by running the heater for 15 minutes every hour during the day. Remember to seal off the unused rooms in your home during a winter blackout as a propane heater is designed to only heat small spaces.

Used conservatively and for small groups, a large propane tank can last for several weeks. Big-box hardware stores sell both the empty propane tanks and heat fixtures. Keep propane stored outside when not in use.

Utilities

If you don't know where your home's utility kill switches are located, spend a few minutes with an experienced family member or neighbor and get acquainted with their location. The last thing you want to be doing is frantically running around the house during a crisis with a flashlight trying to locate the fuse panel or gas line.

Sleeping Bags

The one sleeping bag that my instructors and I have used for many years is the Wiggy's brand. Their design is brilliant, reasonably priced and will outlast any other sleeping bags on the market. Wiggy's uses a continuous layer of insulation rather than the baffling and panels found in other bags so you don't get cold spots, even after years of punishment. When not in use, store them in a non-compressed state in a trash bag.

For ground pads, I've always stuck with Ensolite closed-cell foam pads but the Therm-a-Rest brand is pretty comfy too. The latter require more maintenance (patching) long-term which is why I prefer the closed-cell pads.

Keeping Cool During the Summer Months

While teaching one of my urban survival classes in Phoenix, the question was asked about how to stay cool in the triple-digit heat of a desert city during a long blackout. An older woman in the class relayed her simple advice from a childhood spent in Phoenix before the advent of air-conditioning.

She said that all the homes back then had a screened, enclosed porch off the back of the house. This was lined around the outside with shoulder-high bushes. During the heat of the day she said her mother hosed the bushes and concrete porch down with water for a few minutes. This provided an evaporative cooling effect that lowered the temperature and enabled them to make it through the sweltering afternoons.

At night before bed, her mother would once more spray the bushes and concrete and then the entire family would sleep on cots in the enclosed porch. On extremely hot nights they went a step further and followed the old pioneer trick for staying cool by wrapping up in wet-cotton sheets. It is the same principle as with cold weather situations, only reversed—cool the body not the house. This assumes you have an ample water supply.

You can further augment this time-tested system by purchasing a few battery-operated fans. These can be found in the RV or camping section of an outdoor gear store and there are also some that are solar-powered.

Batteries: how many to stock up on?
We tend to burn through mostly AA and AAA batteries on our camping trips and at home so we keep three dozen of both on hand. 9-volt, C and D batteries aren't in as much demand but we have one dozen of these for various gadgets. While a little pricier, lithium batteries will far outlast standard batteries and I recommend these for critical gear like flashlights and emergency radios.

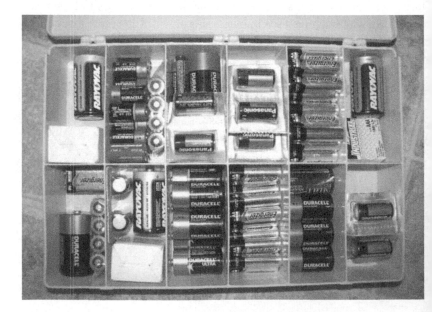

Weather Radios

This is an indispensable item in my line of work for checking on the weather prior to departing on a trip. Get a weather radio from Radio Shack or online that has the NOAA weather band. NOAA (ie. National Weather Service) is where every TV and radio station in the US gets its weather report from so you will be getting information straight from the source for your city. My radio works on either AA batteries or can be recharged using a hand-crank. It also has a USB port and cellphone charger capabilities.

Foldable Solar Panels

These portable solar panels are very low-output in their scope so don't expect it to power up your TV. Whatever items you would normally charge off your vehicle's charger, you can usually charge with one of these panels.

The model I have experience with is the Sunling 12-watt foldable panel which is really only capable of powering gear like 12-volt fans and lights, along with other items you would typically find in the marine or RV industry.

From what a few RVers have told me, another good choice is the GoSolar! 42 Watt Folding Panel as it can power items like cell phones, laptops and video cameras (for filming that epic disaster documentary you've always wanted to make). Keep in mind that the wattage is usually lower than what the manufacturer's specs indicate.

You can extend the longevity of these panels by avoiding repeated folding and not scrunching them down in your pack as this can damage the internal components.

Alternative Fridges

Most Colorado River rafting companies in Arizona use coolers for 14-18 days without any resupply on their trips while in the heat of the desert and we've used them at our basecamp for seven days in triple-digit heat. A couple of tips will help extend the life of food in a cooler:

• Keep the cooler in the shade and on the north side of your house. Pile on some blankets or wet towels to extend the life of the ice inside.
• Bury the cooler in the ground on the north side of house under a protected overhang (so rainwater doesn't leak in). This can be done with or without ice. This method is similar to one used by the pioneers who buried wooden crates in the ground filled with veggies and fruit with a generous layer of straw inside for further insulation. Using this method, we have had carrots and potatoes last for weeks in 80+ degree weather. Secure the lid with duct tape or rocks on top of the lid to prevent critters from getting in.
• Only open the cooler when absolutely necessary.
• Don't drain the ice water out once it has melted as this reduces efficiency. Just make sure all food is secured in Tupperware or Ziplocs or it will turn into slush!

Below is the ice-longevity rate for an 8-gallon cooler that is half-full of food and half covered in ice with an outside air temperature in the '90s during the day.:

• Standard 10 pound bag of ice cubes: 24 hours
• One block of ice: 3 days
• Dry ice: 7 days. When using dry ice, be sure to wear gloves.

DC-Powered Coolers

There are also 12 volt, DC-powered coolers such as the Koolatron that can be plugged into your vehicle's cigarette lighter. I've not used them but many folks in the long-distance trucking community rave about them. Coolers range in size from 26-45 quart.

Alternative Lighting

LED lanterns are the way to go as these will far outlast high-beam and standard light bulbs. Some even come with solar chargers. We also use Coleman propane lanterns at our wilderness basecamps. The latter can really kick out a wide swath of light around the campsite or home but you have to be aware of ventilation issues and potential fire hazards along with needing a constant supply of replacement mantles and propane canisters.

For getting started, I recommend getting one or two of the Coleman 6-volt LED lanterns. These can take either four D batteries or you can purchase the charger. We have used these lanterns on our family camping trips and they last remarkably long compared to the older, battery-powered lanterns.

Regarding candles, the Nu-Wick candles make an excellent addition to both home emergency kits and roadside survival kits for your vehicle. These candles come with multiple wicks to control the heat/light output.

For years, I have used the 44-hour Nu-Wick candles in my truck for heating water on trips, melting snow, and have heard of stranded drivers using these candles to warm the interior of their car (just crack a window!). These candles are non-toxic and unscented and can be used

simply for lighting (with the addition of 1-2 wicks) or heating food (3+ wicks) or making a cup of coffee. They also make 120-hour candles.

Chem-sticks or Cuyalume glo-sticks are handy if you have children and need a temporary nightlight in their rooms. Glo-sticks can provide low-light il-lumination in a house for those times when the power goes out for a few nights. Given this low-output, they should be used to augment your lantern and not relied upon as a primary tool.

Alternative Cooking Methods

If you've already have your favorite backpacking stove, then use it. Other-wise, I recommend a propane stove as it can be used for not only cooking but lighting and heating.

If you don't want a bulbous propane tanker sitting in your backyard, then buy a six-pack of propane canisters. One of these canisters will last for two days of cooking or four nights of lighting a lantern. I've used many camp stoves over the years but have found the tried and true Coleman brands to hold up the best. A two-burner will handle a family of four's cooking needs. You will want to store

he propane outside when not in use rather than in a closet to avoid any fumes
building up.

If you are lucky enough to live near the woods or out in the country, then
campfire cooking is another option. This skill is best learned before you need
it. A 64-ounce stainless steel or enamel cooking pot will work just fine for most
meals or you can use a large coffee can. We have lived for 2-3 weeks on "stews"
cooked in a coffee-can over the fire. Campfire cookery can occupy a whole chap-
er but if you want a good place to start then check out the Boy Scout Manual.
Older editions are better as they cover more traditional campcraft.

Cooking in aluminum foil has been around for ages. A great method for
short-term situations, not so-great for long-term family cooking unless you
have laid in a few hundred rolls of foil. I'd stick to boiling up meals in a 64 ounce
pot or grilling up food on a rack over the hot coals.

I'm a huge fan of Dutch-Oven cooking and own six DOs myself. This is a skill
that takes some trial and error so if you've not done it before do some research
first. Purchase a pre-seasoned Dutch-Oven, and then try your hand at cooking
up a simple dish like a stew, chili, or casserole. Anything that can be cooked in
your home oven can be cooked (even better!) in a Dutch-Oven.

Tips for Buying a Generator

If you are considering purchasing a generator to provide power for your house, first look at what items you will need to run during a disaster. Is it just a few devices like fans or a TV or something larger like a refrigerator or air-conditioner? Lastly, is there anyone in your family who is elderly or has a disability requiring electrical devices? List the items and then tally up the watts required to power them.

You want a generator that is rated over the amount of power you calculated. Then store some gas as you may use 10 to 20 gallons per day! Fuel stabilizer such as Stabil, will need to be added to gas containers when storing fuel for prolonged periods of time. Don't forget to store gas cans in a well-ventilated area outside the home and to run your generator outside away from windows and doors. A battery-powered CO_2 detector inside your house is highly recommended when using a home generator.

As with any pertinent survival gear, give your generator a test-run before you actually need it an emergency. Keep in mind that the noise from a generator will draw unwanted attention to yourself in a long-term situation and gas will become an issue so have a backup plan for supplying power to your house such as some of the alternate sources just listed.

Summary

Alternative Energy and Cooking Starter Kit

6 Nu-Wick candles
2 boxes of Strike-Anywhere matches
6 Lighters
LED Lantern & spare batteries
Two-burner propane stove
Six-pack of propane canisters
Weather radio
24 each AAA and AA batteries
24-quart Cooler
Vehicle cellphone charger
Vehicle power inverter

HANDLING LONG-TERM SANITATION AND HYGIENE ISSUES

It's been day two without power and the toilet is not operational. What will happen after two weeks? The method you use for coping with hygiene and waste management will vary depending on whether you live in a high-rise apartment, the suburbs, or out in the country. Below are a few alternative bathroom solutions that have been employed by families during urban disasters as well as used by the outdoor industry.

Off-Grid Latrines

5-gallon bucket and trash bag

Not much to explain here. Just keep the bag sealed up when not in use and consider using 6 mil contractor bags as they will hold up longer. The bucket-method as with the others that follow are intended for solid waste, not urine.

RV-Style Portable Toilets

There are many versions available and you get what you pay for. These can either be lined with a trash bag or you can use RV chemicals, like Camp-Chem, which is specifically designed to break down human waste and reduce odors.

Outhouse Pit Toilet

The original pioneer method still in use throughout the world today. A 36+" hole is dug and the outhouse is positioned over it. When it is nearing full, move the outhouse a few feet over to another hole, fill the old one, and plant some flowers on the old site. Short of actually constructing an outhouse, you can also use the backpacker's method and dig a cat-hole in the ground at least 4" to 6" deep.

Groovers

These 80mm ammo cans (rocket-boxes) are used by the commercial river-rafting industry in places like the Grand Canyon. Until recently, they were pretty straightforward—with the user sitting on the edge (grooves) of the ammo can to relieve himself. Now there are comfy groovers complete with toilet seats, sealable lid, and an RV-style dumping valve at the bottom for

cleaning. This is the system we use on fieldcourses with the military where we are out for weeks in several different basecamps.

Humanure Method

This is a brilliant, low-tech system for recycling human waste and turning it into compost. It was developed by Joseph Jenkins who has used the method for years and has influenced thousands of homesteaders and off-gridders around the world. To get the scoop, visit his site at http://humanurehandbook.com/

Home Toilet

Many urban survivors have found that they can still use their toilet even when the water main is down. The key is that you will only be flushing it once a day when waste builds up and with water (about a quart) you add in so the unit functions. This is all dependent on whether the nearby infrastructure (plumbing) is still intact so the waste-water has somewhere to go.

Keeping Odors Down

Using a handful of any of the following materials will help to eliminate any foul odors associated with solid waste:

• Sawdust
• Lime (found in hardware stores)
• Wood ash from the fireplace/woodstove/campfire
• Baking Soda
• Peat moss (found in gardening stores)
• Chemicals like bleach or RV chem-treatment fluids
• Cat Litter (hey, it works for them)
• Straw
• Dirt

If the Toilet Paper Runs Out
First off, don't let your TP supply run out- be prepared and stock up. If you have plenty then you will also have bartering material. But worst-case scenario, just look again at what the backpacking community and the pioneers did, in order of preference:
• Non-toxic leaves such as oak, maple, etc… preferably broad-leafed as a clump of pine needles doesn't work well.
• Shredded newspaper/magazines (I can think of a few that should be used for nothing else)
• Juniper, birch or cedar bark (fluff up to soften)
• Smooth river stone(s)
• Snowball

A River Runs Over You: Bathing Methods
During our 9-week programs we used solar camps-showers while at our basecamp and one of these devices will hold enough water for one person. A word of caution: these camp-showers, when left in the afternoon sun, can become scalding hot so I only fill them 2/3 of the way, place in the sun, and then add in 1/3 cold water before using.

During tipi-living days, I just did the sheepherder bath which involved a 5-gallon bucket filled with warm water and a pitcher to dispense. While on the trail, a dip in the lake or river a few times a week was sufficient and we relied on soap obtained from yucca roots or soapwort plants, both of which are found throughout the U.S. Soldiers in the field often do the Helmet Bath in which they use whatever hot water can be placed into their helmet.

A good friend of mine who traveled for extended periods throughout remote villages in Nepal and Tibet, told me of the method used in small settlements that he frequented. Travelers staying at the local "bed and breakfast" were provided each day with a thermos of hot water and a washcloth! This is not a head-to-toe bath and just the essentials were meant to be cleansed. Think of NASA personnel in space on long missions or on the international space labs living for months without a shower. Here, a tri-weekly sponge bath becomes the norm.

When water is scarce and the weather is warm, it can be beneficial to just strip down and sit out in the sun for fifteen minutes as a solar bath is better than none at all.

For what it's worth, I once had to go 21 days without a shower on a long winter survival trip as the temperature was below zero for extended periods. On days when it warmed up (i.e. in the 20 degree F range), I would bathe the essentials using a handful of snow and plenty of expletives. Washing your face, hands and privates is imperative as time goes by in a grid-down situation so don't neglect these regions.

One excellent addition to a home emergency kit, and certainly for a Bail-Out Bag, is to include wet-wipes or baby-wipes. Many of our troops currently rely on the wet-wipe bath for keeping clean in the field long-term. A single package will provide enough wipes for a week of daily cleansing. There is also the No-Rinse shampoos and wipes popular within the RV community.

The Daily Self-Maintenance Routine

We employ the following routine in our long-term survival courses and you will find these beneficial in maintaining hygiene during an extended situation in the city. Plus, these are simple things that make you feel human again after being in the grit and elements.

Wash your face each morning with a damp washcloth or bandanna. You'd be amazed at how this can enhance your PMA.

Brush your teeth. This is easy to forget when you are running on bursts of adrenaline for days on end dealing with the stress of survival.

Change your underwear and socks daily. This is essential in preventing bacteria buildup. Also, women are at greater risk in developing urinary tract infections if they are not changing underwear daily.

Take a full-body sponge bath twice a week, if water permits.

Be aggressive, early on, with treating any minor cuts or scrapes.

Off-Grid Laundry

For use at our basecamp, we've purchased numerous modern off-grid devices billed as tumbling, non-electric washing machines and have found these to not be very durable in the long run. The shell or the handle eventually cracks.

The simplest method, and something I used in younger days while living in a tiny cabin, is the old hippie standby of taking a 5-gallon bucket filled with four gallons of hot water and then applying a plunger to create the intense swirling action needed to clean clothes. Now by clean, I mean the odor and sweat are removed and not necessarily all the stains. This washing method is something that takes place over a vigorous 5-10 minutes of plunging in the bucket so don't expect sparkly new clothes as with an electric washing machine.

One 5-gallon bucket of hot water will take care of a pair of pants, pair of socks, underwear and a T-shirt. A tablespoon of liquid soap will suffice. After rinsing, the clothes are then strung up on a clothesline or hung over tree branches to dry in the sun or wind.

The more humid the environment, the more you will have to wash clothes due to sweat and moisture levels. On desert walkabouts where we are living off the land we have been known to go 14+ days without changing or washing clothes (minus socks and underwear which are washed daily or given solar baths).

Dish-Washing, River-Rafting Style

The best method I've used over the years for long-term sanitation comes from the river-rafting community. On Colorado River trips through the Grand Canyon, where large groups of 30 or more people are out for weeks at a time, hygiene is paramount. Through rigorous testing by generations of professional guides, the four-bucket method was created.

I recommend this method if you are faced with long-term camping in your home or apartment following a disaster. For containers, we use the 3-gallon, galvanized buckets found in hardware stores as these can be placed directly on the stove or campfire to heat the water. If nothing else is available, then pour the water in some large Tupperware, plastic buckets, or even trash bags supported by a ring of rocks on the ground.

River-Rafters' Sparkly-Clean Dishwashing Method

First off, plate-licking is encouraged prior to cleaning. You will want to use every last morsel and to reduce the chunks of food in your dishwashing setup.

Bucket 1- cold or lukewarm water with dishrag for removing any food/chunks
Bucket 2- hot, soapy water with dishrag for wiping plates/bowls
Bucket 3- hot, rinse water
Bucket 4- cold or lukewarm final rinse with bleach. A capful of bleach in the
    bucket will do.

Wipe off dishes or air-dry on a rack. That's it. This is the method we have used for dish-cleaning for weeks on end in the field and it is used by hundreds of river guides each season. The water is replenished in each bucket after every meal. Any food scraps are tossed in the trash, composted or buried in the back-yard, if need be.

Trash Talk

If you have the means of burning trash outside and can safely do so, then this is the route to go in a long-term situation assuming you don't make a plume so big it draws attention to your location.

Newspapers can be saved and used for improvised toilet paper, insulation (crumble it up), firemaking, ground insulation for sleeping, or tied into "logs" and used for firewood in the woodstove or campfire. Cardboard has been used by homeless folks for years for shelter and mats.

Summary

Starter kit for hygiene and dishwashing

5-gallon bucket with a sealable lid
16 extra rolls of toilet paper
Solar camp-shower
2 packages of Wet-wipes per person, per week
12 bars of handsoap
1 gallon of unscented bleach
1 box of 180 trash bags in the 13-gallon size
For infants, a 30-day supply of diapers and pertinent hygiene supplies

Q & A

I kid you not when I say that the following questions have come from my urban survival classes or from media interviews over the years. Some of these were asked with sincerity while others were half-joking.

Question: What should I do to prepare for a potential EMP attack?

Answer: While you are more likely to be adversely affected by a weather-related disaster, EMP attacks are on people's minds today due to the press coverage and sensationalistic movies about their potential impact. I would start by reading the book, *One Second After* and then do some internet research amongst the scientific community. Then I would watch the film *Escape from L.A.* with Kurt Russell who will show you how to prepare for such a crisis with great flare while supplying you with plenty of gritty quotes and manly scowls.

Question: Will my iPod still work if there's an EMP?

Answer: Really dude, get a life. This question is related to a phenomenon I call TMFT or Too Much Free Time.

Question: Is it better to reunite with my family members in one location in my (disaster-torn) city and then bug-out or to just split before the disaster and hook up with them at another location?

Answer: Think of what it's like organizing a family barbecue or birthday party under normal conditions before considering trying to scoop up all of your family members spread across the city while escaping in under 30 minutes. Have plans for reuniting in all four directions with other family members worked out in advance so you can avoid the impending gridlock and make a swift evacuation. I am not saying to abandon grandpa rather stay focused on your immediate family and then concentrate on 1-2 others in your immediate vicinity while having other family similarly responsible for other members in their reach.

Question: Is there really going to be a zombie apocalypse or is this some primal fear of faceless hordes that stems from deep in the limbic part of our ancient reptilian brain?

Answer: Leave zombies alone. They are cool and make for great escapist movies where guys get to run like a wolf pack through a burned-out city and shoot fully automatic weapons with their buddies. Filmmaker George Romero is the Zombie Da Vinci of our times.

Question: Will the pumps at gas stations work if the grid is down for an extended period and how do I extract some without getting busted?

Answer: No, modern fuel pumps will not work when the grid is down as most of these are electronically controlled. In regions with older pumps, then maybe. One sweet, elderly lady in a class of mine, who was a veteran of many hurricanes in the Southeastern U.S., jibed in during a discussion on alternative energy, that "there will be many abandoned vehicles to siphon from and the bottoms of gas tanks puncture easy." When someone asked her if she was suggesting to steal, she replied with a wry smile "File the info away, that's all I'm gonna say on the topic."

Question: I live in Detroit and there are lots of pigeons and sewer rats around. Should I trap and eat these if the grocery shelves empty during a disaster?

Answer: Consider fasting—you will live longer. And get your home pantry in order so you have grub to outlast the disaster recovery phase. Rats in the wilderness taste pretty good but eating rats in the city presupposes you've been vaccinated for plague/rabies/MRSA/plague/kuru/ebola, did I mention plague! You'd be better off crushing gravel and mixing it with tabasco to form a survival gruel.

Question: Can you tell me the <u>exact</u> time, down to the minute, when I should leave the city and bug-out prior to an impending disaster? That's why I came to this class.

Answer: It's almost 4 pm and the class is over after seven hours of lecture, evacuation planning exercises, group scenarios and gear demonstrations. You were in the class today right?!

Question: I own a Hummer, a Jeep, a BMW motorcycle, a Cessna, and a Land Rover and can't decide which one to use in my escape plan. Any suggestions?

Answer: You are suffering from TMFT and the Expendable Cash Syndrome that all too many companies in the survival gear industry cater to. Get a life.

Question: Me and two of my (very wealthy) friends have amassed a ton of food, ammo, firearms, and survival supplies at our cabin retreat in the mountains of northern Montana. The only problem is we all live and work in Arizona. We are pilots, have our own small planes and should be able to get to our retreat if we get out in time but we aren't entirely sure how to use all of our survival gear. What should we do?

Answer: See previous answer.

Question: What do you think will happen if there is ever a global pandemic?

Answer: Good question! Since no modern-day humans have lived through such a pandemic, I can only relay the information and modeling that has been done by epidemiologists and historians based off the 1918-19 Influenza that killed 50 million people worldwide. You can find that material online by looking up the book The Great Influenza: The Story of the Deadliest Pandemic in History by John Barry or by keeping tabs on the work done by the staff at University of Minnesota's CIDRAP research facility—www.cidrap. umn.edu/cidrap/content/influenza/panflu/. Based upon the above information, I think you better have a lot of those Six Priorities in place, a network of like-minded folks in your community, and a whole lot of books in your home library to pass the time as life and commerce will come to a standstill throughout the world for many, many moons.

Question: (A former Marine in an urban survival class once asked me this while I was lecturing on water purification methods) "I just wanna ask…. how many of you could kill a man with your bare hands?"

Answer: You mean before or after you've rehydrated? Maybe we should take a five minute break and clear our heads. Funny thing was that out of a class of 35 people, about 12 hands shot straight up in positive response to the question and those were from the female participants.

RESOURCES

Training Courses and Survival Schools

Urban Skills
The author's Arizona-based company which provides comprehensive survival training courses in urban survival and how to live off the land. Courses range from 1-4 days and are held throughout the U.S. Urban Skills also carries Bail-Out Bags, survival rations, Mora knives, and emergency gear. Visit www.urbanskills.net or call 928-526-2552.

Forager's Path School of Herbal Medicine
Herbalist Mike Masek teaches intensive courses on herbal medicine and edible plants in northern Arizona that range from 2 days to 8-month immersion programs. For more information, visit www.theforagerspath.com.

Ozark Mountain Preparedness
Survival instructor Jerry Ward offers 1-7 day courses in modern survival, trapping, and disaster preparedness at his school in northwest Arkansas. Visit www.ozarkmountainpreparedness.com for more information.

Equipped To Survive
Colleague Doug Ritter's voluminous survival website is the Consumer Reports guide for the survival industry and contains a wealth of gear reviews as well as insightful commentary on previous disasters. Visit www.equipped.org for more information.

Hoods Woods Video Series
Ron and Karen Hood's videos cover everything from making survival kits to trapping and urban survival skills. Karen also publishes Survival Quarterly Magazine. Visit www.survival.com for more information.

Wilderness Medicine Institute
Intensive courses on wilderness medicine ranging from 2-30 days in length as well a good selection of first-aid kits. Visit www.nols.edu/wmi or call 1-866-831-9001.

Suarez International
Gabe Suarez offers some of the finest training courses in self-protection and firearms instruction anywhere. Visit www.suarezinternational.com for his classes or instructional DVDs.

Paxton Quigley

Her books on self-protection and firearms training for women are outstanding. She teaches seminars throughout the U.S. Visit www.paxtonquigley.com for information.

Tom Sotis Combatives

Practical, hands-on combatives courses held throughout the world by renowned instructor Tom Sotis. Visit www.tomsotis.com.

Where to Purchase Quality Gear

Urban Skills

The author's Arizona-based company which provides comprehensive survival training courses in urban survival and how to live off the land. Courses range from 1-4 days and are held throughout the U.S. Urban Skills also carries Bail-Out Bags, survival rations, Mora knives, and emergency gear. Visit www.urbanskills.net or call 928-526-2552.

Adventure Medical Kits (AMK)

AMK has set a new standard for anyone wanting a comprehensive first-aid kit for either individual or group use. Visit www.campmor.com or call 1-800-226-7667.

Scottsdale Gun Club

Full line of firearms, tactical knives, 5.11 gear, and comprehensive firearms and personal protection classes. Visit www.scottsdalegunclub.com

Lehmans

Tons of non-electric products for those seeking greater self-sufficiency and independence. Visit www.lehmans or call 1-888-438-5346.

Nitro-Pak

Supplier of survival kits, dehydrated food, and disaster-related gear. Visit www.nitro-pak.com or call 800-866-4876.

Wiggy's Sleeping Bags

Fine selection of outdoor clothing and the best sleeping bags in the business. Visit www.wiggys.com or call 1-866-411-6465.

Bibliography

ACEP, *Tactical Medicine Essentials,* Jones and Bartlett Learning, 2010

Burns, August A. *Where Women Have No Doctor: A Health Guide For Women,* The Hesperian Foundation, 1997.

Clark, Met. *Emergency Dentistry Handbook.* Paladin Press, 2011.

Forgey, William. *Wilderness Medicine: Beyond Basic First-Aid,* Globe Pequot Pr., 1999 (5th ed.).

Livingston, A.D. *Cold Smoking & Salt-Curing Meat, Fish, & Game,* Globe Pequot Pr., 1995

Nester, Tony. *The Modern Hunter-Gatherer,* Diamond Creek Press, 2009.

Quigley, Paxton. *Armed & Female: Taking Control,* Merril Press, 2010.

Redlener, Irwin. *Americans At Risk: Why We Are Not Prepared For Megadisasters And What We Can Do Now,* Alfred A. Knopf Publishers, 2006.

Ripley, Amanda. *The Unthinkable: Who Survives When Disaster Strikes–and Why.* Three Rivers Press, 2009

Stark, Peter. *Last Breath: Cautionary Tales From the Limits of Human Endurance,* Ballantine Books, 2001.

Suarez, Gabe. *The Tactical Advantage: A Definitive Study of Personal Small-Arms Tactics,* Paladin Press, 1998.

Werner, David. *Where There Is No Doctor: A Village Health Care Handbook.* The Hesperian Foundation, 1992.

Periodicals
Survival Quarterly
Backwoods Home Magazine
Back Home Magazine
Overland Journal
BackWoodsman Magazine

Relevant Films Related To Long-Term Survival
Alive
Alone in the Wilderness (PBS Documentary)
After Armageddon (History Channel DVD)
Defiance
Castaway

If you liked this book, then check out our other titles in the series by Tony Nester:

Practical Survival Tips, Tricks, & Skills

The first book in the series covers how to prepare for emergencies in a forested environment. Topics and skills include: anatomy of survival situations and how to avoid them, lostproofing skills, critical gear every hiker should carry, how to start a fire in wet weather, water purification methods, field-expedient shelters, signal mirror use, and more. Filled with detailed photos and practical tips.

$10.95 plus $4 shipping. Available at Apathways.com or Amazon.com.

Desert Survival Tips, Tricks, & Skills

The second book in the series focuses on the skills and strategies for adapting to arid regions. Topics covered include: survival psychology, clothing selection, heat-related injuries and how to avoid them, shade shelters, the latest treatment methods for venomous bites and stings, water location tips, myths regarding water consumption in the desert, preparing kids for desert trips, and how to equip your vehicle. Jammed with photos and pragmatic skills. This book is utilized by the US Military.

$10.95 plus $4 shipping. Available at Apathways.com or Amazon com.

Surviving A Disaster:Evacuation Strategies and Emergency Kits For Staying Alive

In the third book in the Practical Survival Series, Tony Nester takes you through the scenarios, planning, and emergency kits for surviving natural and manmade disasters where you are forced to evacuate your home. Surviving A Disaster covers what has worked for real-life survivors and delves into the practical skills that you can use to prepare your family.

$10.95 plus $4 shipping. Available at Apathways.com or Amazon.com and Kindle.

The Modern Hunter-Gatherer: A Practical Guide to Living Off the Land

In his fourth book, survival instructor Tony Nester delves into practical methods that he has applied on extended survival courses over the past twenty years showing the best techniques for beginning and advanced students of wilderness living. This innovative book illustrates, with detailed photos, the essential methods for harvesting, preserving, and cooking small game, fish, edible plants, and how to reduce your dependence on "the system."

Cost: $16.95 plus $4 shipping. Available at Apathways.com or Amazon.com and Kindle.

Tony Nester's Practical Urban Survival DVD Series, Volumes 1 & 2

This instructional DVD series covers the skills, strategies, and essential gear to have in place to help you prevail in an urban crisis at home, work, or on the road. The methods and skills shown in both DVDs flow in a logical order and cover how to prepare for both short-term emergencies ranging from 1-3 days to a long-term grid-down situation lasting several weeks or longer. The material is delivered in a straightforward, no-frills fashion with solid information that can be immediately applied to your home and lifestyle. Visit www.apathways.com.

About the Author

Tony Nester is the founder of the highly-respected survival school, Ancient Pathways in Flagstaff, Arizona. His company is the primary provider of survival training for Military Special Operations units from around the world and he has instructed the National Transportation & Safety Board (NTSB), FAA, and served as a technical consultant for the film Into the Wild. Tony is the author of four previous books and three survival DVDs. He has been featured on Fox  News, the Discovery Channel, Travel Channel, the *New York Times*, *Backpacker* and *Maxim Magazine*. He and his family live in a passive-solar, strawbale house in northern Arizona.

For more information, contact:

Tony Nester
Ancient Pathways, LLC
PO Box 2068
Flagstaff, AZ 86003
928-526-2552
www.urbanskills.net
www.apathways.com

Made in the USA
Las Vegas, NV
13 February 2024

85618609R00046